Ayodeji Ijagbuji, I. I. Zakharov, T. C. Philips, M. G. Loriya, M. B. Saltzberg, A. B. Tselishtev, R. J. Taylor, B.V. Shevchenko, K. M. Jackson, D. B. Woods, J. K. Johnson

Production of olefins via oxidative de-hydrogenation of C3 C4 fraction by O2 over (Cr Mo)SiO2

GRIN Publishing

Bibliographic information published by the German National Library:

The German National Library lists this publication in the National Bibliography;
detailed bibliographic data are available on the Internet at http://dnb.dnb.de .

Imprint:

Copyright © 2015 GRIN Verlag GmbH
Print and binding: Books on Demand GmbH, Norderstedt Germany
ISBN: 978-3-656-90003-0

This book at GRIN:

http://www.grin.com/en/e-book/292808/production-of-olefins-via-oxidative-de-
hydrogenation-of-c3-c4-fraction

GRIN - Your knowledge has value

Since its foundation in 1998, GRIN has specialized in publishing academic texts by students, college teachers and other academics as e-book and printed book. The website www.grin.com is an ideal platform for presenting term papers, final papers, scientific essays, dissertations and specialist books.

Visit us on the internet:

http://www.grin.com/

http://www.facebook.com/grincom

http://www.twitter.com/grin_com

Production of olefins via oxidative de-hydrogenation of C_3–C_4 fraction by O_2 over (Cr–Mo)/SiO_2

A. A. Ijagbuji[1*], I. I. Zakharov[1,2], T. C. Philips[1], M. G. Loriya[1], M. B. Saltzberg[3], A. B. Tselishtev[1], R. J. Taylor[4], B.V. Shevchenko[3], K. M. Jackson[4], D. B. Woods[4], J. K. Johnson[4]

[1]*Institute of Technology, East Ukrainian National University*, Severodonetsk, 93400, Ukraine
[2]*Boreskov Institute of Catalysis*, Siberia, Russia
[3] *Moscow State University*, Moscow, Russia
[4] *University of Melbourne*, Parkville, Victoria, Australia

Abstract

The present study investigates the oxidative dehydrogenation of propane-butane (C_3–C_4) fraction over mono (Cr or Mo) and bi-metal (Cr-Mo) loaded SiO_2 catalysts. The catalysts were prepared by sequential impregnation method at 500°C calcination temperature. Experiments were performed by feeding C_3–C_4 fraction, oxygen, nitrogen, and steam into a continuous flow quartz reactor at an atmospheric pressure (P = 1 atm.), reaction temperatures between $500 – 650^\circ$C, gas hourly space velocity (GHSV) within $100 – 400$ h^{-1}, and at reaction time (t_r) = 2h. Appropriate water vapor addition to the feed sinificantly minimizes oxidation into coke deposits on the catalyst surface, and also prevents further olefin conversion into undesirable product. The physicochemical properties were evaluated by BET, XRD, IR, and EPR characterization techniques. The major oxidation products are *ethylene, propylene, isobutylene, butylene*. This paper reports that the total yield of olefins (Σ C_2-C_4) = 66.0 % was achieved at 83.5 % conversion level of C_3–C_4 at 630°C. The results indicate that the addition of Mo to catalysts of Cr/SiO_2 modifies its catalytic activity for the ODH reaction. Mono-metallic catalysts (Cr/SiO_2 and Mo/SiO_2) were prepared for comparison purposes.

Keywords: oxidative de-hydrogenation of C_3–C_4, C_3–C_4 conversion, selectivity to olefin, olefin yield, olefin production.

1. Introduction

Olefins are the most important starting materials for modern industrial organic chemistry. Over the past several decades, the manufacturing capacities and importance of low molecular-weight olefin (*ethylene, propylene, isobutylene, and butylene*) have undergone remarkable transitions, as well as continue to serve as a fundamental basis for the petrochemical and refining industries. The catalytic de-hydrogenation of alkanes has a considerable industrial impact because it represents a route to obtain olefins from feed-stocks of low-cost saturated hydrocarbons.[1]

Traditional technologies for olefins production involve catalytic dehydrogenation of alkanes via either the steam cracking or fluid catalytic cracking method. Dehydrogenation reactions appear very simple: their thermodynamic and kinetic characteristics have, nevertheless, contributed to make the development of technologies that allow for a reliable and efficient industrial application, rather complex i.e. the overall efficiency of olefin production by the catalytic dehydrogenation method is greatly impeded by a few drawbacks such as: (i) thermodynamic limitations, owing to the highly endothermic and the substantial energy requirement to activate the reactant molecules; (ii) coke deposits build up on the reactor wall thus, lowering heat transfer, increasing pressure and corrosion.[2] Consequently, the existing commercial processes for olefins production *are very energy-intensive, exhaustive, and not cost-effective*. While these two routes are very well developed, increasing the capacity of these processes is only possible to some extent, as changing regulation limits the use of by-products (*notably aromatic molecules*) in fuels. For these reasons, the present industrial capacity for C_2-C_4 olefins production via these traditional processes is expected to be insufficient, and therefore, cannot meet the fast-growing demand of olefins in the international market.[3]

Since producers seek to leverage their existing assets and the available internal streams to find an optimum solution for meeting the demands of olefins, oxidative de-hydrogenation, which involves coupling the reagent mixture with oxidant such as oxygen [Eq. (1) – (2)], has been widely studied, as a potentially attractive route to circumvent the thermodynamic limitations, eliminate coking, and therefore, extend catalyst lifetime.

$$C_3H_8 + {}^1/_2O_2 \rightarrow C_3H_6 + H_2O \qquad \Delta H_f^0 = -95.5 \ kJ/mol \qquad (1)$$
$$C_4H_{10} + {}^1/_2O_2 \rightarrow C_4H_8 + H_2O \qquad \Delta H_f^0 = -95.5 \ kJ/mol \qquad (2)$$

There are, however, a number of current challenges preventing oxidative de-hydrogenation from being widely implemented. The difficulties inherent in oxidative dehydrogenation reactions revolve around selectivity control because all equivalent C–H bonds have an equal bonding energy, and therefore an equal chance of reacting.[4] When two C–H bonds of neighboring carbons are split, a double bond is formed and alkanes are converted to alkenes. Thereby, oxygen addition to alkane feeds exposes the synthesized olefins to further oxidation conditions that results into the formation of *environmentally-damaging and economically-useless carbon oxides* (CO and CO_2), consequently decreasing the yield of alkenes [Eq. (3) – (4)].

$$C_3H_6 + 3O_2 \rightarrow 3CO + 3H_2O \qquad \Delta H_f^0 = -219 \ kJ/mol \qquad (3a)$$
$$C_3H_6 + {}^9/_2O_2 \rightarrow 3CO_2 + 3H_2O \qquad \Delta H_f^0 = -1867.5 \ kJ/mol \qquad (3b)$$
$$C_4H_8 + 4O_2 \rightarrow 4CO + 4H_2O \qquad \Delta H_f^0 = -518 \ kJ/mol \qquad (4a)$$
$$C_4H_8 + 6O_2 \rightarrow 4CO_2 + 4H_2O \qquad \Delta H_f^0 = -2716 \ kJ/mol \qquad (4b)$$

Therefore, the design of effective catalytic systems that are sufficiently active, exhibit high selectivity, be periodically-regenerated under severe conditions, and yet operate at temperatures that minimize oxygenation of the desired products, are key performance demands for cost-effective production of olefins. Nevertheless, the mechanism of carbon filament formation and catalyst de-activation resulting from the decomposition of hydrocarbons on catalyst metal particles has been extensively studied in the past.[5 – 11] Thereby, if the cracking of lower alkanes is to be utilized for olefin production in a continuous process, the addition of steam (*as a diluent*) at 550°C has been demonstrated to restore the catalytic activity even after complete catalyst deactivation, and enhance higher equilibrium conversion.

During the past decade, the design and synthesis of bi-metallic catalysts have attracted considerable attention because they show multiple functionalities and prominent catalytic activity, selectivity, and stability over monometallic catalyst[12 – 16] i.e. bi-metallic catalytic systems can achieve chemical transformations that are unprecedented with because different components of the catalyst have a particular function in the overall reaction mechanism.[17 – 19] For instance, the use of chromium-based catalyst enhances the presence of amorphous silica dioxide phase, whereas molybdenum exhibits excellent catalyst attrition resistance, facilitates easy products desorption from the catalyst surface, maintain optical defect concentration, and blocks non-selective sites. Thereby, the new physical and chemical properties derived from synergistic effects between the two metals are highly desirable for catalytic applications. Incorporation of hetero-atoms into a silica framework has been reported to increase thermal stability [20] and may also increase the acidity of the support. [21]

Transition metal oxides have a tremendous importance in the field of heterogeneous catalysis, serving as either catalysts or as supports for other catalytically active species. For instance, amorphous silica is an important support in catalytic technology due to its thermal stability, tunable porosity, outstanding specific surface area, and tremendous metal(s)-support interaction. [19, 22 – 24] Generally, it is believed that for metals on an irreducible oxide support: the strength of metal-metal bond is significantly larger than that of the metal-support bond. In addition, depending on the particular metal-oxide system, oxidation and reduction at elevated temperatures are essential steps for the preparation of high surface area supported catalysts; notwithstanding, these treatments can cause various morphological alterations such as *thermal-sintering*,[25 – 26] *encapsulation*,[27 – 28] *inter-diffusion*,[29 – 31] and *alloy formation*.[32] In particular, *alloy formation* from metals supported on silica has received considerable attention because of its adverse influence to significantly alter catalytic activity and selectivity.[33 – 34]

A unique feature of supported metal oxide catalysts is that the active component should be exclusively present as a surface phase, 100 % dispersed, below monolayer coverage, and that there is no spectroscopic complication from the co-existence of bulk crystalline phases. In addition, It is generally accepted that the activation of a hydrocarbon molecule on oxide catalysts takes place on a centre involving a M–O acid–base couple, oxide ions (O): playing the role of a hydrogen atom abstraction centre, cationic centers (M): facilitating the electron transfer. However, in spite of the numerous studies[19 – 34] on metals supported on SiO_2 at elevated temperatures, there are still controversial and unresolved issues regarding: (i) the nature of the metal-support interaction between metals and SiO_2; (ii) the morphological changes that occur during the high temperature reduction of metals supported on SiO_2; (iii) the role of oxygen vacancies in the inter-diffusion of metals into SiO_2; (iv) the extent to which silicides are formed by the direct interaction between metals and SiO_2; (v) the role of the silicon substrate, frequently used to prepare SiO_2 thin films, in metal silicide formation; (vi) the composition of metal silicides (if formed); (vii) the mechanism of silicide formation the between metals and SiO_2. Notwithstanding, since catalyst activity and selectivity[35 – 37] are highly dependent on the size, shape, and nature of oxide support, it is therefore, of considerable importance to investigate and define the optimal conditions for catalyst preparation, pre-treatment and activation.[38] An atomistic understanding of catalyst systems is essential to the rational design of improved catalyst systems.

To the best of our knowledge, there has been only one report in the literature, so far, describing the synthesis of bi-functional Cr-Mo/SiO_2 catalyst and its application for the catalytic oxidative desulfurization of diesel fuel.[39]

The objective of this contribution was to verify the morphological alterations after catalyst synthesis, and study the effects of Cr, Mo, and (Cr-Mo) supported silica catalysts on the oxidative dehydrogenation of C_3–C_4 fraction to olefin, at temperatures between 500 – 650°C, atmospheric pressure (P = 1 atm.), gas hourly space velocity (GHSV) within 100 – 400 h^{-1}, and at reaction time (t_r) = 2h in a continuous flow quartz reactor.

2. Experimental

2.1 Materials and Methods:

Tetraethyl orthosilicate (TEOS, 99 %) was obtained from *Sigma-Aldrich*. Ethanol (C_2H_5OH, 99.5 %) was obtained from *Acros Organics*. Citric acid ($C_6H_8O_7$) and hydrochloric acid (HCl, 37 %) were obtained from *Fischer Scientific*. Propane-butane fraction ($C_3–C_4$, 99.9 %) was supplied by *Naftogas*, Kiev. Oxygen (O_2, 99.99 %) and Nitrogen (N_2, 99.99 %) were obtained from *Azot chemical company*; and distilled water (H_2O, 1 dm^3).

2.2 Catalysts Preparation:

Wet impregnation was used in the preparation of co-impregnated catalysts. The quantitative metal precursors (Cr, Mo, or Cr–Mo) were dissolved in 15 mL de-ionized water. The pH of the aqueous solution was adjusted to 2–3 by adding 0.5 mol/L *citric acid*. 21 mL of *tetraethyl orthosilicate* (TEOS) and 22 mL of *ethanol* (C_2H_5OH) were added to the solution to produce the TEOS-C_2H_5OH solution. Then, the aqueous metallic salt solution was dropped slowly into the alcohol solution, and the mixed solution was stirred at room temperature for 1 h to produce sol. The sol was placed at room temperature for 1 h and aged at 40°C for 1 h to produce gel. The gel was dried under ambient atmosphere at 120°C for 12 h, and the catalyst precursor was obtained. Finally, the precursor was calcined at 500°C for 4 h. The amorphous silica (*surface area* = 200 m^2/g) was carefully treated by washing in 2 M of HCl (*hydrochloric acid*) to remove any volatile impurities adsorbed on the surface (**Fig. 2**). (i) The 10 %wt. Cr mono-metallic catalyst was prepared by incipient wetting of amorphous silica with an aqueous solution of chromium chloride ($CrCl_2$); (ii) The 15 *wt*.% Mo mono-metallic catalyst was prepared by incipient wetting of amorphous silica with an aqueous solution of ammonium heptamolybdate ($[NH_4]_6Mo7O_{24}.6H_2O$); and (iii) The 15 *wt*.% Cr-Mo bi-metallic catalyst was prepared by slowly dropping aqueous solutions of chromium chloride and ammonium heptamolybdate ($[NH_4]_6Mo7O_{24}.6H_2O$) in a shaking water-bath at 20°C. After the impregnation, the resulting mixture was stirred, filtered, and then dried at 150°C. Then both mono and bi-metallic catalyst sample were pressed, and then sieved to appropriate sizes for catalytic evaluations. Metal(s) loading was varied between 10 – 15 %wt.

2.3 Catalysts Treatment:

0.1g of each catalyst was calcined in air at 500°C for 6 h to remove any volatile impurity adsorbed on the surface, followed by reduction in 10 % H_2/ 90% Ar at 500°C for 6 h to minimize coking and enhance its dehydrogenation activity without influencing the secondary reaction of light olefins production. The total flow rate of the H_2/Ar mixture was 50cm^3/min. The samples were then cooled at room temperature for 30mins, and stored in inert atmosphere to avoid catalyst degradation.

2.4 Catalysts Test:

The catalytic activity of the samples for the decomposition of $C_3–C_4$ fraction was investigated in a continuous flow quartz fixed-bed reactor (6 L vol.). The catalysts sample (0.15 g) was packed in the reactor and activated with flowing N_2 at 500°C for 2 h. After which the flow rate of reactant ($C_3–C_4$) and N_2 was maintained at 4.0 ml/min and 96 ml/min, respectively. The N_2 adsorption isotherms of calcined materials were measured at liquid nitrogen temperature ($-196°C$). The gas carrier was passed through a molecular sieve trap before being saturated with C_3–C_4. The gas product samples were analyzed by gas chromatograph.

2.5 Oxidative De-hydrogenation of C_3–C_4:

The C_3–C_4 feed composition was analyzed by gas chromatography «Chrom–5». It was established that the total composition equals 100 % by *volume*: propane = 20, *i*-butane = 60, and *n*-butane = 20. The oxidative dehydrogenation experiment of C_3–C_4 fraction was carried out in a continuous flow quartz fixed-bed reactor (6 L vol.). The experimental set up is shown in Fig. 1.

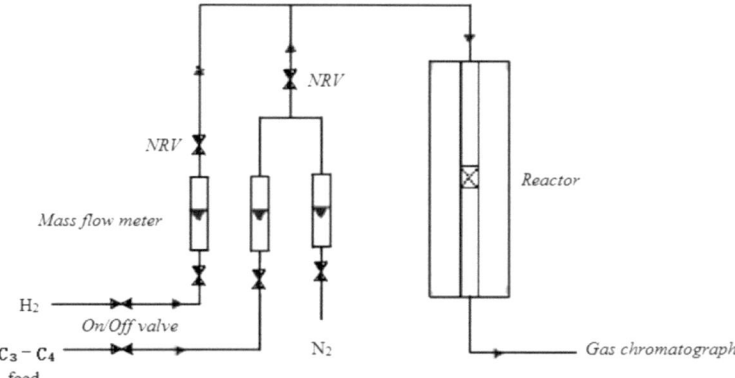

Fig. 1   *Schematic diagram of the pilot unit for dehydrogenation of C_3–C_4 fraction*

Prior to each run, the reactor was purged with N_2 for about 10 min and then de-coked using 15 % O_2: 85 % N_2 mixture to ensure that the reactor walls and the coupon were coke free. This was accomplished by visually observing the appearance of the coupon through an observation hole in the furnace and by monitoring weight of the coupon during the decoking process. If the appearance of the coupon was transparent and non-luminous, and its weight did not decrease with time, the coupon was assumed to be coke free. The reactor was again purged with N_2 for about 10 min, after which the hydrocarbon reactants and steam were introduced. The primary reason for N_2 purge before and after decoking experiments was to minimize the accumulation of potentially explosive mixtures in the reactor. Each run was repeated at least five times to ensure reproducibility and to assess the range of experimental errors associated with the experiments.

In order to determine the catalytic specie with the best performance, three series of experiment were performed. In the 1st *experiment*, C_3-C_4–oxygen–vapor mixture on the 10 %wt. Cr/SiO$_2$ catalyst was fed into the reactor. The propane-butane fraction, oxygen, and nitrogen were supplied from a pressure cylinder through reduction valves. The reactant gases consisting of C_3–C_4, O_2, and some additional N_2 carrier gas were then mixed with steam and transported to the reactor through electrically heated lines at a flow-rate desired ($\pm$ 5%) for the given experiment. A molecular sieve was used at the entrance of the reactor with the objective of retaining impurities coming from the feeding line. The flow rate of C_3–C_4 fraction, O_2, and N_2 was regulated by mass flow controller that was calibrated before the experiments. The reactant mixture was subjected to thermal treatment at temperature range within T = 500 – 650°C, atmospheric pressure (P = 1 *atm.*), and for (t_r) = 2 h total reaction time. The flow of bulk gas through the reactor was 100 cm^3/min, and 1 L/min for water. The average residence time of reactant mixture in the reactor was about 10 s. The gas hourly space velocity of 100 – 400 h^{-1} was varied to obtain different level of C_3–C_4 conversion. The catalyst activity was maintained by regeneration after every experimental hour using a N_2/O_2 mixture. During the regeneration, the output stream from the reactor was checked for carbon dioxide. The reaction products were obtained at the bottom of reactor, cooled, separated into individual components, and analyzed at 10 min. intervals by gas chromatograph–mass spectrometer (GC–MS).

In the 2nd *experiment*, C_3-C_4–oxygen–vapor mixture on the 15 %*wt.* Mo/SiO$_2$ catalyst was fed into the reactor. In the 3rd *experiment*, C_3-C_4–oxygen–vapor mixture on 15 %*wt.* (Cr-Mo)/SiO$_2$ catalyst was fed into the reactor. The same experimental condition described in 1st experiment was applied for the 2nd and 3rd experiments.

3. Results and Discussion

3.1 *Catalysts Characterization*:

The Textural characteristics such as metallic composition, BET surface area, and pore distribution of catalyst samples are compiled in Table 1. Atomic absorption spectroscopy by «Accu-sorb» was used to measure the elemental composition of catalysts. The BET surface areas (determined by N_2 physisorption) and the pore size distributions of all the catalyst samples were calculated using the *Brunauer-Emmett-Teller* (BET) and *Barrett-Joyner-Halenda* (BJH) methods, respectively. Upon Cr, Mo, and Cr-Mo incorporation on the silica support, the surface areas and the pore volumes were reduced considerably compared to that of silica support (200 m^2/g). From the data (Table 1), it can be seen that the surface areas and pore volumes of these samples show a similar trend; being maximum for 15 %*wt.* Cr-Mo/SiO$_2$ (178.6 m^2/g, 0.452 cm^3/g), intermediate for the 10 %*wt.* Cr/SiO$_2$ (164.3 m^2/g, 0.439 cm^3/g), and minimum for the 15 %*wt.* Mo/SiO$_2$ (161.5 m^2/g, 0.376 cm^3/g).

A clear bimodal metal particle size distribution of SiO$_2$ thin film was analyzed using scanning electron microscopy «EM–125K». It was observed that particle size of catalysts has an insignificant effect on C_3–C_4 conversion, thus, the effect of internal mass transfer resistance may be neglected. The 15 %*wt.* Mo/SiO$_2$ catalyst exhibits intermediate particle sizes of diameter(s) d = 5 – 10 *nm*, the 15 %*wt.* Cr-Mo/SiO$_2$ catalyst shows large particle sizes of diameter(s) d = 50 – 100 *nm*, whereas, the 10 %*wt.* Cr/SiO$_2$ catalyst possesses a very large particle size of diameter $d \geq 200$ *nm* (Table 1).

Table 1   *Physicochemical properties of the different catalyst samples*

Catalysts	Composition (wt.%)		Surface area (m^2/g)	Pore volume, V_p (cm^3/g)	Pore diameter, d (*nm*)
	Cr	Mo			
10 %*wt.* Cr/SiO$_2$	10	–	161.5	0.376	> 200
15 %*wt.* Mo/SiO$_2$	–	15	164.3	0.439	5 – 10
15 %*wt.* Cr-Mo/SiO$_2$	10	5	178.6	0.452	50 – 100

The slight reductions in textural characterization are due to the presence of large amount of Cr, Mo, and Cr-Mo species outside the framework of silica. In addition, the differences shown by the mono and bi-metallic catalysts are often related to different conditions that the dissolved complex of Cr and Mo encounter during the impregnation process i.e. on one hand, the aqueous solution of $CrCl_2$ behaves as a strong acid (with a pH around 2.5), whereas, $([NH_4]_6Mo_7O_{24}.6H_2O)$ aqueous solution shows a weak acid character (with a pH around 5.6); on the other hand, the mixture of aqueous solutions of $CrCl_2$ and $([NH_4]_6Mo_7O_{24}.6H_2O)$ produces bimetallic co-impregnated catalysts with an average pH of 4.3. It has been established that at pH values between 4.0 to 5.0, a hydrolytic reaction occurs in $CrCl_2$ solution, to form colloidal $Cr(OH)_2$ species. At this pH value, the molybdate anions $(Mo7O_{24}^{6-})$ continue being the dominant species during impregnation process. The simultaneous interaction of Cr and Mo species with the positively electrically-charged surface of silica propitiates the selective adsorption of $Mo7O_{24}^{6-}$ anions on the surface, followed by precipitation of Cr particles during the evaporation of the aqueous solution. The values of pore diameter of samples are much greater than the molecular diameter of water (2.75 Å or 0.275 nm); hence, the effects of diffusive retardation and re-adsorption inside pores are suppressed to a maximum extent.

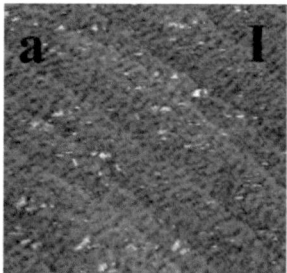 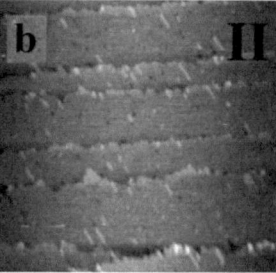

Fig. 2 SEM micrographs of: (a) clean SiO_2 thin film (*considered to be the more defective surface*); (b) clean SiO_2 thin film (*considered to be the less defective surface*).

3.2 x-ray diffraction of calcined catalysts:

The crystalline phases in samples of each catalyst was studied using X-ray diffractometer (XRD). XRD can be particularly useful in catalysis because specific crystalline phases may be active for specific types of reactions. The X-ray tube was operated at 40 kV and its scanning rate was $5°$/min. The K_α radiation of diffracted beam mono-chromator was selected and angular range from $5°$ to $35°$ was recorded using step scanning. The result indicates slight variations in crystal structures of 10 %wt. Cr/SiO_2, 15 %wt. Mo/SiO_2, and 15 %wt. $Cr-Mo/SiO_2$. The three compounds showed well resolved four diffraction lines at 2θ range = 7.5, 15, 23.3 and 28 (Fig. 3).

Sintering can be understood using the *Gibbs–Thompson* relationship where larger particles with lower chemical potential will grow at the expense of smaller particles with higher chemical potential: the driving force being the reduction of the total surface energy of the system. With these considerations, metals with an electro-negativity $\chi < 1.5$ on the Pauling scale should react with a SiO_2 substrate. Therefore, sintering was not observed in both mono-functional catalysts, owing to the relatively high electro-negativity values of χ Cr = 1.66 and χ Mo = 2.16.

It has been revealed in study [33 – 34] that the formation of alloy can occur by direct interaction between silica and the metal, and it is generally believed that oxygen vacancies are the primary diffusion pathways. Although, a strong interaction does exist between each mono-metal (Cr or Mo) and the SiO_2 support, the high temperature reduction via diffusion of metal through the silica layer significantly lowers the barrier to silicon diffusion, results to point defects on SiO_2 thin films, thus, preferentially leading to alloy formation. In this study, X-ray data revealed the formation of transition-metal silicide species such as chromium silicide (e.g. Cr_2SiO_4, Cr_3Si, and $CrSi_2$) in the 10 %wt. Cr/SiO_2, and also molybdenum silicide (e.g. $MoSi_2$ and Cr_3Si) in the 15 %wt. Mo/SiO_2. In contrast to alloy formation in both mono-metallic catalysts, no alloy feature was observed in the 15 %wt. $Cr-Mo/SiO_2$. This may be due to the bi-metals (Cr-Mo) and the support (Si) nucleating separately, and thereby, not mixing below the desorption temperature of Si.

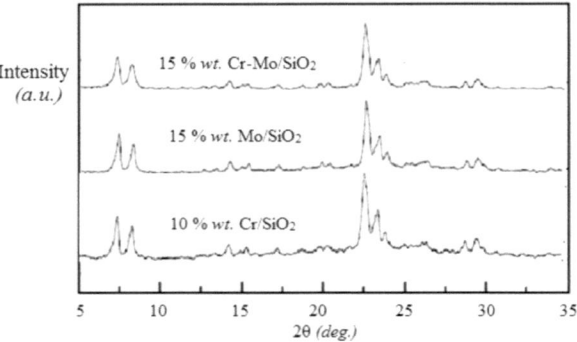

Fig. 3   XRD data of the calcined catalysts

3.3 NH_3 adsorption sites of catalysts:

Since, the irreversible NH_3 adsorption implies localized chemi-sorption of single NH_3 molecules on the acidic sites, this quantity provides the total number of acidic sites on a solid surface. The numbers of surface acidic sites on the catalysts are shown in Table 2. Of the three catalyst samples investigated in the study, the Cr-Mo impregnated SiO_2 sample (15 %wt. $Cr-Mo/SiO_2$) possesses the greatest number of acidic sites per unit surface area of adsorbent. The order of amount of NH_3 consumed in catalysts is as follows: 15 %wt. $Cr-Mo/SiO_2$ > 10 %wt. Cr/SiO_2 > 10 %wt. Mo/SiO_2. This increase in weak acid sites with molybdenum addition could probably be due to the increased concentration of surface-isolated tetrahedral Mo species.

Table 2   Number of surface acidic sites on catalysts by NH_3 adsorption at 500°C and P = 1 atm.

Catalysts	Surface area, (m^2/g)	Number of acid sites $(mmolNH_3/g)$	Number of acid sites $(\mu mol/m^2)$
10 %wt. Cr/SiO_2	164.3	0.20	1.22
15 %wt. Mo/SiO_2	161.5	0.09	0.06
15 %wt. $Cr-Mo/SiO_2$	178.6	0.32	1.79

Particularly, the 15 %wt. Cr-Mo/SiO$_2$ showed a high total acidity indicating the presence of large fraction of exposed active Mo sites. However, to obtain robust control of stereo-chemistry, a fine balance should be struck between the *number of acid sites* and *acid strength of* catalysts. The strength of acid sites (*surface hydroxyl group*) on the catalysts were analyzed via pyridine adsorption using IR spectroscopy «Specord IR–75» in the frequency range of 500 – 4000 cm^{-1}. A strong and wide absorption band, caused by the stretching vibration of the O–H bond, appears at 3450 cm^{-1} (Fig. 4). However, a weak absorption band appears at 1639 cm^{-1} due to the flexural vibration of the O–H bond. Absorbed water and hydroxyl on the sample surface contribute to these absorption bands. A strong absorption band, caused by vibration of the chemical bond of the surface siloxane group (Si–O–Si), is present at approximately 1089 cm^{-1}. In the 10 % wt. Cr/SiO$_2$ sample, the vibration of the chemical bond of Si–O–Si moves toward the low wave velocity and causes a strong absorption band at approximately 1083 cm^{-1}. This condition also indicates a certain interaction between chromium and the supporting matrix.

From the IR spectrum (Fig. 4), the vibration of the chemical bond of Si–O–Si causes a strong absorption band at approximately 1089 cm^{-1}. However, in the sample, a strong absorption band is observed at around 1080 cm^{-1} when the vibration of the chemical bond of Si–O–Si moves toward the low wave velocity. This phenomenon also indicates the effect of Mo addition. Moreover, despite the addition of Cr and Mo, the absorption band still presents a slight variation in the infrared spectrum as a result of the insufficient amount of Mo. The order of acid intensities of the catalyst samples follow the sequence: 15 %wt. (Cr-Mo)/SiO$_2$ > 10 %wt. Cr/SiO$_2$ > 15 %wt. Mo/SiO$_2$.

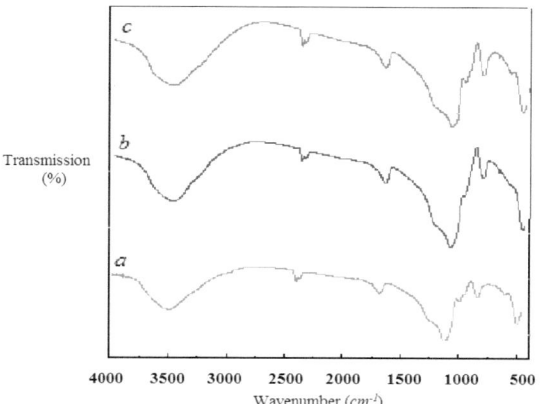

Fig. 4 *IR Spectra of Catalysts*: (a) 10 %wt. Cr/SiO$_2$; (b) 15 %wt. Mo/SiO$_2$; (c) 15 %wt. Cr-Mo/SiO$_2$

Owing to the complex surface chemistry of Cr, the oxidation state of active species has been the subject of debate and controversy for many years. According to several authors, the active species in alkane dehydrogenation is Cr^{3+}, while other authors propose both Cr^{2+} and Cr^{3+}, or solely Cr^{2+} as the active species. In addition to that, Cr^{5+} ions and even Cr$_2$O$_3$ clusters can, also, be formed on Cr/SiO$_2$ catalysts. These differences also explain why the data in the literature are sometimes difficult to compare and only by a combination of different spectroscopic techniques can detailed molecular-level information be obtained. [40 – 41]

The oxidation state of the active species and interaction of the metal(s) with an oxide carrier was studied by X-ray photoelectron spectroscopy (XPS). The XPS spectroscopic analysis in this work confirmed the presence of Cr^{3+}, Cr^{5+}, and Mo^{5+} in the 15 %wt. Cr-Mo/SiO$_2$ catalyst. It is suggested that Cr^{6+} in tetrahedral coordination are formed as an active mono-chromate species and reduced to Cr^{3+} in octahedral coordination as a less active poly-chromate species during the reaction. Deactivated catalyst was regenerated by a treatment with gaseous oxygen, during which a re-oxidation of the Cr^{3+} species to Cr^{6+} species was observed. The reduced octahedrally coordinated $Cr(III)O_6$ can be re-oxidized to the tetrahedrally coordinated $Cr(VI)O_4$ by oxygen. Therefore, the reduction–oxidation cycle between the tetrahedral symmetry of $Cr(VI)O_4$ and octahedral symmetry of $Cr(III)O_8$ plays an important role in the dehydrogenation of C_3–C_4 with oxygen over SiO_2 i.e. when the 15 %wt. Cr-Mo/SiO$_2$ catalyst was preheated at 500°C for 1 h, and then subsequently treated with C_3–C_4; a broad β–signal located at proportionality factor (*g-factor*) ≈ 1.94, which is attributed to the Cr^{3+} ion begins to appear. With an increase of temperature to 630°C, the intensity of β-signal reaches a maximum. As the temperature is further increased to 650°C, no changes were observed in the intensity of β–signal. The 15 %wt. Cr-Mo/SiO$_2$ catalyst possesses the largest concentration of paramagnetic centers (*conventional units* = 165).

3.4 *Oxidative de-hydrogenation of C_3–C_4:*

During oxidation, the alkane molecule is adsorbed to the surface of oxygen atom [Eq. (5)], followed by the cleavage of C–H bond to form an alkyl intermediate, and a hydroxyl group on the catalyst surface [Eq. (6)]. The adsorbed alkyl specie looses a second hydrogen atom, thus, forming alkene, and another hydroxyl group on the catalyst surface [Eq. (7)]. Two hydroxyl groups combine to form water and lattice vacancy [Eq. (8)]. The oxidative dehydrogenation reaction is postulated to occur via a redox cycle, where the catalysts lattice oxygen takes part in the oxidation reaction [Eq. (9)], and then the reduced catalyst is re-oxidized following a *Mars-van Krevelen* mechanism.[45]

$$C_nH_{2n+2} + O^* \leftrightarrow C_nH_{2n+2}O^* \quad (n = 3, 4) \tag{5}$$
$$C_3H_8 + O^* \leftrightarrow C_3H_8O^* \tag{5a}$$
$$C_4H_{10} + O^* \leftrightarrow C_4H_{10}O^* \tag{5b}$$
$$C_nH_{2n+2}O^* + O^* \rightarrow C_nH_{2n+1}O^* + OH^* \quad (n = 3, 4) \tag{6}$$
$$C_3H_8O^* + O^* \rightarrow C_3H_7O^* + OH^* \tag{6a}$$
$$C_4H_{10}O^* + O^* \rightarrow C_4H_9O^* + OH^* \tag{6b}$$
$$C_nH_{2n+1}O^* \rightarrow C_nH_{2n} + OH^* \quad (n = 3, 4) \tag{7}$$
$$C_3H_7O^* \rightarrow C_3H_6 + OH^* \tag{7a}$$
$$C_4H_9O^* \rightarrow C_4H_8 + OH^* \tag{7b}$$
$$2OH^* \leftrightarrow H_2O + V^* + O^* \tag{8}$$
$$O_2 + 2V^* \rightarrow 2O^* \tag{9}$$

The catalytic performances of samples, as well as the effect of external mass transfer on C_3–C_4 oxidation was studied by varying stirrer speed from 100 – 400 h^{-1}, T = 500 – 650°C, and P = 1 atm. Since the pore structure of catalyst surface vary at different reaction temperatures, any impurity generated at different temperatures blocks the catalyst pore structure, resulting in lower catalytic activity and shorter catalyst effective time. Therefore, five different temperatures ranging from 500 to 650°C were selected to investigate the optimum reaction temperature. The optimum reaction temperature was therefore, carefully determined by CO$_x$ removal efficiency and effective time of catalyst. The best experimental result was achieved at a gas hourly space velocity of 250 h^{-1}, which is presented in Table 3. *Gas hourly space velocity is the gas speed during a reaction and affects various catalytic reactions differently* i.e. a low *gas hourly space velocity* is more suitable for catalysts with a dense structure, whereas high *gas hourly space velocity* is effective for catalysts with a loose structure. In this study, five values of *gas hourly space velocity* at 100 h^{-1}, 200 h^{-1}, 250 h^{-1}, 300 h^{-1}, and 400 h^{-1} were used to evaluate the effects on C$_3$–C$_4$ *conversion*.

Table 3   *Typical catalytic performances in propane-butane oxidative conversion*

Catalysts	T (°C)	Conv. (%) C$_3$-C$_4$	Selectivity (%) C$_2$H$_4$	C$_3$H$_6$	iC$_4$H$_8$	CH$_4$	C$_2$H$_6$	Yield (%) C$_2$H$_4$	C$_3$H$_6$	iC$_4$H$_8$	CH$_4$	C$_2$H$_6$
Cr/SiO$_2$												
	500	30.9	20.9	17.1	14.0	–	–	6.5	5.3	4.3	–	–
	550	47.3	25.2	16.0	12.3	–	–	11.9	7.6	5.8	–	–
	600	59.9	27.3	13.2	9.0	–	–	16.4	7.9	5.4	–	–
	630	63.6	35.0	9.8	7.1	–	–	22.3	6.2	4.5	–	–
	650	68.4	42.8	7.2	4.7	27.5	13.8	29.3	4.9	3.2	12.5	6.3
Mo/SiO$_2$												
	500	21.3	14.8	12.9	10.6	–	–	3.2	2.7	2.3	–	–
	550	33.7	18.0	11.2	8.9	–	–	6.1	3.8	3.0	–	–
	600	41.1	19.7	9.5	7.1	–	–	8.1	3.9	2.9	–	–
	630	43.8	25.6	6.9	5.5	–	–	11.2	3.0	2.4	–	–
	650	47.5	29.3	5.2	3.8	23.8	10.6	13.9	2.5	1.8	11.3	5.0
Cr-Mo/SiO$_2$												
	500	44.7	38.2	36.9	19.2	–	–	17.1	16.5	8.6	–	–
	550	58.6	39.4	34.5	15.3	–	–	23.1	20.2	9.0	–	–
	600	69.6	41.5	32.8	12.9	–	–	28.9	22.8	9.0	–	–
	630	83.5	43.5	25.6	9.8	–	–	39.0	22.7	8.8	–	–
	650	89.6	46.2	23.3	2.3	18.9	9.3	41.0	20.7	2.0	16.8	8.3

P = 1 atm., GHSV = 250 h^{-1}, C$_3$–C$_4$: O$_2$: H$_2$O = 1 : 1 : 3, and reaction time (t$_r$) = 2h.

The feed conversion (X), olefin selectivity (S) and olefin yield (Y) were calculated using following relationship:

$$Conversion\ (\%) = \frac{(\text{content of C}_3\text{-C}_4\ \text{feed}) - (\text{content of C}_3\text{-C}_4\ \text{in product})}{\text{content of C}_3\text{-C}_4\ \text{feed}}$$

$$Selectivity\ of\ olefins\ (\%) = \frac{(\%\ \text{content of olefins in product}) \times 100}{\text{C}_3\text{–C}_4\ \text{conversion}\ (\%)}$$

$$Yield\ of\ olefins\ (\%) = (\%\ \text{C}_3\text{–C}_4\ \text{conversion}) \times (\%\ \text{olefin selectivity})$$

Table 3 shows that when the *gas hourly space velocity* = 250 h^{-1}, the C$_3$–C$_4$ oxidation reached a conversion level of 89.6 % and the effective time was 10 min. With increased *gas hourly space velocity* from 250 h^{-1} to 400 h^{-1}, the *oxidative conversion* rate of the catalyst significantly decreased. The removal rate also decreased when *gas hourly space velocity* = 100 h^{-1}. Thus, the optimum GHSV was determined to be 250 h^{-1}. When the *gas hourly space velocity* was increased, the C$_3$–C$_4$

oxidation efficiency increases due to decrease of the out-diffusion. When the g*as hourly space velocity* was further increased, the *conversion level* increases correspondingly, suggesting that an increase of g*as hourly space velocity* resulted in increase in gas conversion but shortened retention time of gas, thus reaction tends to be incomplete. From this point of view, the proper g*as hourly space velocity* was of great importance for *propane-butane oxidative de-hydrogenation.*

From the data presented in Table 3: (i) When the reaction temperature was varied from 500 to 650°C on the 10 *wt*.% Cr/SiO_2 catalyst, it can be seen that the selectivity to olefin slightly increases from 52.0 to 54.7 % as C_3-C_4 conversion level, and the total yield of olefins (Σ C_2-C_4) increase from 30.9 to 68.4 %., and 16.1 to 37.4 %, respectively; (ii) When the reaction temperature was varied from 500 to 650°C on the 10 *wt*.% Mo/SiO_2 catalyst, no significant increment in olefin selectivity was observed. Nevertheless, C_3-C_4 conversion level increases from 21.3 to 47.5 %., as does the total yield of olefins (Σ C_2-C_4) from 8.2 to 18.2 %; (iii) When the reaction temperature was varied from 500 to 650°C on the 10 *wt*. % $(Cr-Mo)/SiO_2$ catalyst, the C_3-C_4 conversion level increases from 44.7 to 89.6 %, as does the total olefins yield (Σ C_2-C_4) from 42.2 to 66.0 % (at 630°C), followed by a sharp reduction to 63.7 % at 650°C.

It can be seen from this result that the C_3-C_4 conversion level and selectivity to ethylene formation increases with an increasing temperature. On the contrary, selectivity to propylene and butylenes decreases with increase in both reaction temperature and C_3-C_4 conversion level. The corresponding decrease in C_3-C_4 alkenes selectivity at high C_3-C_4 conversion level may be attributed to higher reactivity of C_3 and C_4 olefins than the reactant mixture (C_3-C_4 with oxygen) over the catalysts. It has been reported that activation energy requirements for dehydrogenation decreases with the increase in carbon chain. [46] For instance, butane is expected to be more quickly cracked over bi-metallic silica support comparative to propane i.e. while butane is partially cracked to either propylene and methane ($C_4H_{10} \rightarrow C_3H_6 + CH_4$) or to ethylene and ethane ($C_4H_{10} \rightarrow C_2H_4 + C_2H_6$); propane; is partially cracked to ethylene and methane ($C_3H_8 \rightarrow C_2H_4 + CH_4$). In addition to that, propylene molecule (C_3H_6) has the tendency to form allylic site (delocalized unpaired electron) or weak C-H links that are liable to secondary reaction, consequently, decreasing its yield. As a result, it is not feasible to get higher yield or selectivity to propylene/butylene at increased temperature. Nevertheless, at higher conversion level of C_3–C_4, the particularly low selectivity to C_4 alkenes is due to hydride transfer reaction activation, and also suggests C_4 alkanes to be the dominant reaction in the investigated system. Moreover, it was observed that increase in the reaction temperature beyond 630°C has undesirable effect on olefin selectivity i.e. the formation of cracking products such as C_1, and C_2-hydrocarbons were observed at more pronounced quantities from the reaction products. It can, therefore, be concluded from the data obtained in this study that ethylene selectivity is maximized by increasing the reaction temperature, either by preheating the reactants or by using oxygen enriched air, whereas propylene and butylenes selectivity are maximized by lowering the reaction temperature.

The effect of mole ratio of water vapor on the conversion of C_3–C_4 was studied by varying the mole ratio from C_3-C_4–H_2O = 1: 3 to 1: 9 respectively, but keeping all other reaction parameters to be constant. It was observed that increasing the ratio of vapor mixture from 3 to 9 (*%mol.*), practically, has no effect on the feed conversion and/or olefins yield. This may be attributed to absence of oxygenates in reaction products.

The effects of 10 *%wt*. Cr/SiO_2, 15 *%wt*. Mo/SiO_2, and 15 *%wt*. $Cr-Mo/SiO_2$ catalysts on the oxidative de-hydrogenation of C_3–C_4 to olefins were investigated (Table 2). It can be seen that the 15 *%wt*. Mo/SiO_2 exhibits a low catalytic activity, selectivity, and yield. *The results from the activity test for the oxidative dehydrogenation reaction of C_3–C_4 fraction showed an increment in the* *specific activity towards olefin formation when Mo was added.* Upon the addition of 5 %wt. Mo to 10 %wt. Cr/SiO$_2$, a two-fold conversion level of C$_3$–C$_4$ (83.5 %), relative to the mono-functional catalysts, and total yield of olefins: ΣC$_2$-C$_4$ = 66.0 %, were achieved. Hence, the overall yield of olefins over 15 %wt. Cr-Mo/SiO$_2$ catalyst, apparently gets an enhancement in comparison to values obtained for mono-metallic catalysts (10 %wt. Cr/SiO$_2$ and 15 %wt. Mo/SiO$_2$).

The scanning electron microscopy (SEM) analysis of the 15 %wt. Cr-Mo/SiO$_2$ (not shown) after multiple reaction cycles, shows that the carbonization degree results to sub-metallic particle formation «*cliftonite*» which, in contrast to graphite, is not only an easily removable form of coke, but also enhances the catalyst self-regeneration upon incorporation of water vapor. The steam addition significantly minimizes carburization on catalyst surface, and also prevents further olefin conversion to aromatic compounds (favored by high temperatures). The results show that most of the filamentous carbon is removed during steam regeneration, leaving only small pockets of this material which resist this treatment. The absence of the Cr-Mo alloy and solid-solution formation implies no alteration in chemical structure of the active material; this fact, along with the virtual catalytic inactivity of Mo/SiO$_2$ (Table 2) lead to the conclusion that the improved specific activity, is as a result of interaction of the MoO$_x$ species and Cr crystallites. This interaction could be related to a modification in the mechanisms of reaction in which the MoO$_x$ might facilitate insertion of oxygen into alkane molecules, therefore, avoiding the oxidation of C$_3$–C$_4$ to CO, CO$_2$, or other oxygenated hydrocarbon compounds. The enhancement of surface acidity of silanol groups and the higher facile reducibility as a result of strong interaction between Cr-Mo and SiO$_2$ support may be the reason for improved results of C$_3$–C$_4$ conversion and olefins yield over the bi-metallic supported catalysts (15 %wt. Cr-Mo/SiO$_2$). Therefore, from the XRD result in this work (Fig. 3) for the oxidative de-hydrogenation of C$_3$–C$_4$ over transition metal oxide supported catalyst: our perspective on reducibility as a key factor for catalyst activity agrees with study.[47]

When the catalytic results (Table 3) is correlated with the surface and structural characteristics (Table 1 and Fig. 3, respectively), it is of interest to note that the decrease sequences of C$_3$–C$_4$ conversion over the catalysts is identical to that of the total peak areas of catalysts. This result indicates that the reduction peak has a close relationship with the oxidative de-hydrogenation reaction and suggests that the species giving rise to this peak facilitates olefins production.

At reaction temperatures between T = 500 – 630°C, in-situ spectroscopic analysis of the reaction products was performed using gas Chromatograph «Chrom–5». The major oxidation products are *ethylene, propylene, isobutylene,* and *water.* In addition to that, neither cracking products (such as C$_1$, C$_2$-hydrocarbons), aromatic compounds (e.g. C$_6$H$_6$), nor oxygenated compounds (e.g. R–CHO or CH$_3$OH) were present in the exhaust of the reactor. Since C$_3$ and C$_4$ olefins are destroyed through subsequent reactions (secondary cracking), it is likely that an optimized plant design, with strong integration between reaction and cooling zones, will reduce the incidence of this undesired effect. Though, the partial contribution to olefins via heterogeneous de-hydrogenation or oxidative dehydrogenation reactions cannot be excluded, the strong dependence of all hydrocarbon products on feedstock conversion (directly related to the operating temperature), is a clear indication that homogeneous reactions plays a major role.[43] Also, from a technical view-point, our perspective that a strong interplay between hetero-homogeneous and exo-endothermic reactions, a proper choice of active phase, as well as catalytic reactor design are of paramount importance for achieving and maintaining high catalytic performance, is consistent with the literature.[48]

The catalytic evaluation shows that Cr content of 10wt.% and Mo content of 5 wt.% over SiO$_2$ designated as 10Cr–5Mo/SiO$_2$ shows maximal result in both olefins selectivity (78.9 %), and total yield of olefins (Σ C$_2$-C$_4$) = 66.0 % at C$_3$–C$_4$ conversion level (83.5 %) under reaction temperature T

$= 630°C$, $P = 1$ atm., and gas hourly space velocity $= 250$ h^{-1}. Under the mild treatment, the catalyst was tested for two consecutive runs, and the catalyst activity was found to be stable. Thereby, our observation on the tremendous catalytic activity of $(Cr-Mo)/SiO_2$ in oxidative de-hydrogenation reaction is broadly supportive of the previous study. [39]

Nevertheless, it is imperative to clearly state that the maximum overall yield of olefins is dependent on a number of parameters such as catalysts pre-treatment/preparation, the type of support, the nature and amount of oxidant used, as well as the operating conditions (e.g. reaction temperature, regeneration–de-hydrogenation cycles, reactant flow-rate, etc.).

4. Conclusion

The oxidative dehydrogenation of propane-butane fraction to olefins in the presence of oxygen over mono (Cr or Mo) and bi-metal (Cr-Mo) loaded SiO_2 catalysts, at temperatures $T = 500 - 650°C$, atmospheric pressure ($P = 1$ atm.), and gas hourly space velocity (GHSV) $= 100 - 400$ h^{-1}, has been investigated in a continuous flow quartz reactor.

$$2(C_3H_8 + C_4H_{10}) + 2O_2 + 6H_2O \rightarrow 2C_2H_4 + 2C_3H_6 + C_4H_8 + 10H_2O \qquad \Delta H_f^0 = -36kJ/mol \qquad (10)$$

From this study, it can be concluded that:

(i) The oxidative dehydrogenation at a molar ratio of C_3–C_4 to oxygen $= 1$, $T = 630°C$, $P = 1$ atm., and gas hourly space velocity (GHSV) $= 250$ h^{-1} over $(Cr–Mo)/SiO_2$ catalyst are optimal conditions for obtaining the highest selectivity (78.9 %) and yield of olefins (ΣC_2-$C_4 = 66.0$ %);
(ii) The 15 %*wt.* $Cr-Mo/SiO_2$ catalyst showed a better tolerance to carbonaceous deposits, in comparison to the mono-metallic catalysts (Cr/SiO_2 and Mo/SiO_2);
(iii) The presence of Mo enhances the catalytic performance of $Cr-Mo/SiO_2$ for the selective oxidative de-hydrogenation of C_3–C_4;
(iv) The $(Cr-Mo)/SiO_2$ exhibits the most preferred catalytic performance at calcination temperature of $500°C$;
(v) The catalytic efficiency on C_3–C_4 conversion, selectivity to olefins, and olefins yield are in the sequence: 15 %*wt.* $Cr-Mo/SiO_2 \gg$ 10 %*wt.* $Cr/SiO_2 \gg$ 15 %*wt.* Mo/SiO_2;
(vi) The raw materials (e.g. C_3–C_4, O_2, and H_2O) are inexpensive and abundantly available, thus, making the overall production process to be cost-effective.

Conclusively, the oxidative de-hydrogenation of C_3–C_4 fraction in the presence of air and steam (*gas–vapor system*) is an environmentally benign technology and can therefore, be considered as a useful guidance to design an industrial plan for the conversion of low molecular weight paraffin hydrocarbons to olefins; however, this novel technology will be ready for the industrial application only after completing an extensive experimental work in the range of operating conditions identified as the most promising by technical-economic evaluations, and after improving the catalyst and reactor design reliability with respect to heat management.

Acknowledgements

This work was supported by the *State Foundation for Basic Research*. The co-operations of *Prof.* B. C. Scott and *Prof.* C. W. Anderson are gratefully acknowledged.

References

[1] B.M. Weckhuysen, I.E. Wachs, R.A. Schoonheydt, *Chem. Rev.* 96 (1996) 3327.

[2] Chan, K.Y.G., F. Inal, and S. Senkan, Suppression of coke formation in the steam cracking of alkanes: ethane and propane, *Industrial & Engineering Chemistry Research*, 1998 37 (3) 901 – 907.

[3] Corma, A.; Melo, F.V.; Sauvanaud, L.; Ortega, F. *Catal. Today,* 699 (2005) 107 – 108.

[4] J.E. Germain, *Catalytic Conversion of Hydrocarbons*, Academic Press, London, 1969.

[5] R.T.K. Baker, M.A. Barber, P.S. Harris, F.S. Feates, R.J. Waite, *J. Catal.* 26 (1972) 51.

[6] R.T.K. Baker, P.S. Harris, R.B. Thomas, R.J. Waite, *J. Catal.* 30 (1973) 86.

[7] J. Rostrup-Nielsen, *J. Catal.*, 48 (1977) 155.

[8] R.T.K. Baker, *Catal. Rev. Sci. Eng.* 19 (1979) 161.

[9] R.T. Yang, J.P. Chen, *J. Catal.* 115 (1989) 52.

[10] N.M. Rodriguez, *J. Mater. Res.* 8 (1993) 3233.

[11] R.T.K. Baker, M.S. Kim, A. Chambers, C. Park, N.M. Rodriguez, *Catalyst Deactivation*, 99 (1997)

[12] V.R. Stamenkovic, B. Fowler, B.S. Mun, G. Wang, P.N. Ross, C.A. Lucas, N.M. Markovic, *Science*, 315 (2007) 493.

[13] F. Tao, M.E. Grass, Y. Zhang, D.R. Butcher, J.R. Renzas, Z. Liu, J.Y. Chung, B.S. Mun, M. Salmeron, G.A. Somorjai, *Science*, 322 (2008) 932.

[14] B. Lim, M. Jiang, P.H.C. Camargo, E.C. Cho, J. Tao, X. Lu, Y. Zhu, Y. Xia, *Science*, 324 (2009) 1302.

[15] T. Omori, K. Ando, M. Okano, X. Xu, Y. Tanaka, I. Ohnuma, R. Kainuma, K. Ishida, *Science*, 333 (2011) 68.

[16] E. González, J. Arbiol, V.F. Puntes, *Science*, 334, (2011), 1377.

[17] M. Chen, D. Kumar, C.W. Yi, D.W. Goodman, *Science*, 310, (2005), 291.

[18] L. Kesavan, R. Tiruvalam, M.H. Ab Rahim, M.I. bin Saiman, D.I. Enache, R.L. Jenkins, N. Dimitratos, J.A. Lopez-Sanchez, S.H. Taylor, D.W. Knight, C.J. Kiely, G.J. Hutchings, *Science*, 331 (2011) 195.

[19] G. Kyriakou, M.B. Boucher, A.D. Jewell, E.A. Lewis, T.J. Lawton, A.E. Baber, H.L. Tierney, M. Flytzani-Stephanopoulos, E.C.H. Sykes, *Science*, 335 (2012) 1209.

[20] M.T. Bore, T.L. Ward, R.F. Marzke, A.K. Datye, *J. Mater. Chem*, 15 (2005) 5022 – 5028.

[21] R. Ravishankar, M.M. Li, A. Borgna, *Catalysis Today* 106 (2005) 149 – 153.

[22] A. Katrib, C. Petit, P. Legare, L. Hilaire, G. Maire, *Surf. Sci.*, 886 (1987), 189 – 190.

[23] R. Lamber, N. Jaeger, G. Schulz-Ekloff, *Surf. Sci.*, 227, (1990), 268

[24] R. Lamber, W. Romanowski, *J. Catal*, 105 (1987), 213.

[25] M. Chen, L.D. Schmidt, *J. Catal.*, 55, (1978), 348.

[26] X. Xu, D.W. Goodman, *Appl. Phys. Lett.*, 61, (1992), 1799.

[27] B.R. Powell, S.E. Whittington, *J. Catal.*, 81, (1983), 382.

[28] L.C.A. Van den Oetelaar, A. Partridge, S.L.G. Toussaint, C.F.J. Flipse, H.H. Brongersma, *J. Phys. Chem. B*, 102 (1998) 9541.

[29] R. Anton, U. Neukirch, M. Harsdorff, *Phys. Chem. B*, 36, (1987) 7422.

[30] H. Dallaporta, M. Liehr, J.E. Lewis, *Phys. Chem. B*, 41, (1990), 5075.

[31] M. Liehr, H. Dallaporta, J.E. Lewis, *Appl. Phys. Lett.*, 53, (1988), 589.

[32] L.C.A. Van den Oetelaar, R.J.A. Van den Oetelaar, A. Partridge, C.F.J. Flipse, H.H. Brongersma, *Appl. Phys. Lett.*, 74, (1999), 2954.

[33] L.L. Shen, Z. Karpinksi, W.M.H. Sachtler, *J. Phys. Chem.* 93 (1989) 4890

[34] W. Juszczyk, Z. Karpinski, *J. Catal.* 117 (1989) 519

[35] S.A. Stevenson, R.T.K. Baker, J.A. Dumesic, E. Ruckenstein (Eds.), Metal-Support Interactions in Catalysis, Sintering, and Re-dispersion, *Catalysis Series*, Van Nostrand Reinhold, New York, 1987.

[36] M. Valden, X. Lai, D.W. Goodman, *Science*, 281 (1998) 5383.

[37] M. Haruta, *Catal. Today* 36 (1997) 153

[38] T.C. Chang, J.J. Chen, C.T. Yeh, *J. Catal.*, 96, (1985), 51

[39] X. Jiang-ping, L. Qina, W. Ning, Zhang Yuqin, Zhang Liyu, Ma Yali, Qiu Jianghua, Wang Guanghui, Synthesis of Cr-Mo/SiO$_2$ and its application in catalytic oxidative desulfurization of diesel fuel, *Journal of Wuhan University of Science and Technology* (2011) Issue 4, 280 – 284.

[40] M. Liehr, H. Lefakis, F.K. Legoues, G.W. Rubloff, *Phys. Rev. B,* 33 (1986) 5517.

[41] B.M. Weckhuysen, I.E. Wachs, R.A. Schoonheydt, *Chem. Rev.* 96 (1996) 3327.

[42] D.E. O'Reilly, D.S. MacIver, *J. Phys. Chem.* 66 (1962) 276.

[43] D.E. O'Reilly, F.D. Santiago, R.G. Squires, *J. Phys. Chem.* 73(1969) 3172.

[44] D. Cordischi, M.C. Campa, V. Indovina, M. Occhiuzzi, *J. Chem. Soc., Faraday Trans.* 90(1) (1994) 207.

[45] G. Karamullaoglu, T. Dogu, *Ind and Engr. Chem. Res.* 46 (2007) 7079 – 7086.

[46] R. Serge, *Thermal and Catalytic Processes in Petroleum Refining*, Marcel Dekker, New York, 2003.

[47] H.X. Dai, A.T. Bell, E. Iglesia, *Journal of Catalysis* 221 (2004) 491 – 499.

[48] M. Huff, L.D. Schmidt, Production of olefins by oxidative de-hydrogenation of propane and butane over monoliths at short contact times, *J. Catal.* 149 (1994) 127 – 141.